NOSTALGIAWORKS
COMPANION READING

AF879879

COPYRIGHT AND PERMISSIONS NOTICE

For partnership inquiries, bulk orders, institutional sales, or permissions request, contact NostalgiaWorks at info@nostalgiaworks.com or visit nostalglaworks.com

NANA'S BOOKS, a Connecticut Limited Liability Company,
DBA NostalgiaWorks
Laurette Klier © 2020

We have just begun to fight!
PEARL HARBOR
BATAAN
CORAL SEA
MIDWAY
GUADALCANAL
NEW GUINEA
BISMARCK SEA
CASABLANCA
ALGIERS
TUNISIA

Brave men rejoice in adversity,
just as brave soldiers triumph
in war.

-Lucius Annaeus Seneca

Man the
GUNS
Join the NAVY

Americans never quit.

-General Douglas MacArthur

Honor the Brav
1861-1865
Copyright 1907 by Illustrated Postal Card Co

The American flag represents all of us and all the values we hold sacred.

-Adrian Cronauer

7th
WAR LOAN
NOW.. ALL TOGETHER

One flag, one land, one heart
one hand, one nation evermore!

-Oliver Wendell Holmes

Build
for your NAVY!
ENLIST
CARPENTERS, MACHINISTS, ELECTRICIANS ETC.
FOR INFORMATION APPLY TO YOUR NEAREST RECRUITING STATION
U.S. NAVY – BUREAU OF YARDS & DOCKS

Any man who may be asked what he
did to make his life worthwhile,
I think can respond with a good deal
of pride and satisfaction:

'I served in the United States Navy'.

-John F. Kennedy

35
38
TEAMWORK WINS
UNITED STATES SHIPPING BOARD EMERGENCY FLEET CORPORATION

Do not pray for easier lives.
Pray to be stronger men.

-John F. Kennedy

U.S.
NAVY

A good Navy is not a provocation to war, it is the surest guaranty of peace.

-Theodore Roosevelt

WE FOUGHT FOR
YOUR RIGHT TO VOTE
PLEASE USE IT

The ballot is stronger than the bullet.

-Abraham Lincoln

"THE MARINES HAVE LANDED"!
APPLY TO NEAREST RECRUITING STATION

Only our individual faith in freedom can keep us free.

-Dwight D. Eisenhower

This will remain the land of the free, so long as it is the home of the brave.

-Elmer Davis

"Sub Spotted—
Let 'Em Have It!"

Fear is a reaction.
Courage is a decision.

-Winston Churchill

Keep him flying!
BUY WAR BONDS

Those who stand for nothing
will fall for anything.

-Alexander Hamilton

"above and beyond the call of duty"

What lies behind us and what lies before us are tiny matters compared to what lies within us.

-Unknown

Memorial Day
Souvenir

If your actions inspire others to dream more, learn more, do more, and become more, you are a leader.

-John Quincy Adams

A thoughtful mind, when it sees a nation's flag, sees not the flag, but the nation itself.

-Henry Ward Beecher

OVER THE TOP
FOR YOU
Buy U.S. Gov't Bonds
THIRD LIBERTY LOAN

Fortune favors the brave.

-Terence

How lucky I am to have someone who makes saying goodbye so hard.

-A.A. Milne

We Can Do It!
POST FEB. 15 TO FEB. 28
WAR PRODUCTION CO-ORDINATING COMMITTEE

The highest obligation and privilege of citizenship is that of bearing arms for one's country.

-George S. Patton, Jr.

FIRST IN FRANCE
U.S. MARINES

Guard zealously your right to serve in the Armed Forces, for without them, there would be no other rights to guard.

-John F. Kennedy

YOUR VICTORY GARDEN
counts more than ever!

I think patriotism is like charity,
it begins at home.

-Henry James

U.S. MARINES
"SOLDIERS OF THE SEA"
BRUCE MOORE

A ship without Marines is like
a garment without buttons.

-Admiral David D. Porter

GO HIGHER
FARTHER
FASTER
U.S. AIR FORCE

I have long believed that sacrifice is the pinnacle of patriotism.

-Bob Riley

BETTINA-
STEINKE

It is foolish and wrong to mourn the men who died.

Rather, we should thank God that such men lived.

-George S. Patton, Jr.

ARMY
U.S.
"KEEP 'EM FLYING"
IS OUR BATTLE CRY!
. . .
DO YOUR PART
FOR
DUTY - HONOR - COUNTRY
. . .

There is something reassuring about a man who will willingly ascend from the heavens solely for the purpose of setting others free.

-Anonymous

PAUL STAHR

Liberty is the breath of life to Nations.

-George Bernard Shaw

BOOK CHAT

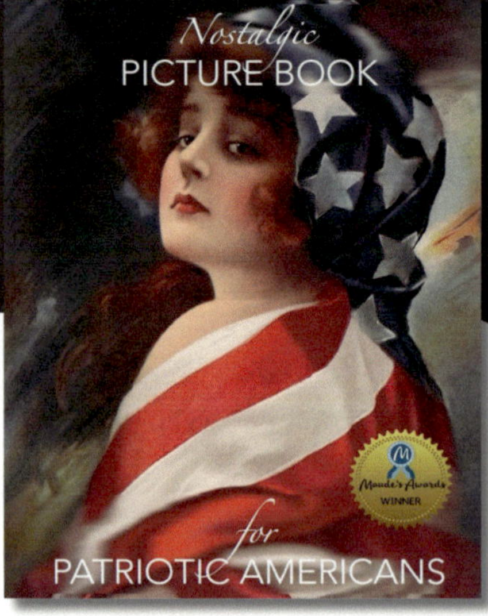

LET'S FIND...

-Lily of the valley flowers
-A monkey
-A palm tree
-A fighter jet
-Two potatoes
-Shoes with spats
-A polka-dot kerchief

FUN FACTS!

-The US Army has the most active duty personnel.
-The US Coast Guard has the fewest active duty personnel.

-The Army has had the most presidents serve overall, though since WWII, most veteran presidents served in the Navy.

-On average, only one in four applicants is eligible to serve.
-Established in 1775, the Army is the oldest branch of service.
-The Veterans of Foreign Wars (VFW) was the first to sell red poppies in memory of the fallen.
-Memorial Day fell on May 30 up until 1971.

AUTHOR SHARE:

I am a Memorial Day baby, born on May 30, 1967.
Perhaps that explains my deep gratitude for military families because I celebrated my birthday on the day we honored their service and sacrifices.

Made in the USA
Las Vegas, NV
10 November 2026

57851949R00033